Cover images taken from Our Changing Planet: The U.S. Climate Change Science Program for the Fiscal Year 2007. Prepared by the U.S. Climate Change Science Program. (http://www.globalchange.gov/resources/gallery?func=viewcate gory&catid=22)

NCA Report Series, Volume 5a

Ecosystem Responses to Climate Change: Selecting Indicators and Integrating Observational Networks

NCA Report Series

The NCA Report Series summarizes regional, sectoral, and process-related workshops and discussions being held as a part of the Third National Climate Assessment (NCA) process. Three workshops focusing on monitoring changes in the physical climate system, as well as the impacts of climate change on ecosystems and socio-economic systems were completed in support of the 2013 NCA and the ongoing NCA process. The first of these workshops, on developing indicators for ecosystem responses to climate change and integrating observational networks, was organized by the Ecosystem Interagency Working Group of the U.S. Global Change Research Program. The workshop was held in Washington, DC on November 30-December 1, 2010 and sponsored by the U.S. Environmental Protection Agency. Volume 5a of the NCA Report Series summarizes the discussions and outcomes of this workshop. A list of planned and completed reports in the NCA Report Series can be found online at http://assessment.globalchange.gov.

CONTENTS

CONTENTS

EXECUTIVE SUMMARY

Nearly 60 representatives from across the ecological monitoring community attended the workshop Ecosystem Responses to Climate Change: Selecting Indicators and Integrating Monitoring Systems hosted by the Ecosystems Interagency Working Group (EIWG) on November 30 and December 1, 2010. This group was convened to make progress toward two goals:

- Outline a process for selecting indicators that represent the impacts of climate change on the nation's ecosystems. The intended use of this set of indicators is the ongoing assessment of impacts of climate change. These indicators would be incorporated in the National Climate Assessment (NCA), including in the next report, to be issued in 2013.

- Identify opportunities for collaboration and coordination among existing and potential future observational networks that could be used to improve the understanding of the impacts of climate change on the nation's ecosystems.

Through a series of plenary presentations, plenary discussions, and breakout group discussions, several **key issues** emerged regarding these two goals. Participants raised several important points that differentiated past efforts to select indicators, and in many cases, were related to the ultimate efficacy of the indicators as research tools, guides for decision-making, or communication devices. Several of these points also addressed the extent and types of information that indicators could (and could not) provide about the impacts of climate change on ecosystems. These points included:

- Establishing the intended purpose of a set of indicators;

- Using indicators to understand attribution or cause-and-effect relationships;

- Tying indicators to a conceptual model of ecological function and ecosystem services;

- Designating the parties involved with the indicator selection process; and

- Achieving buy-in from users and establishing the indicators' association with management decisions.

The participants also discussed past and ongoing efforts to integrate, or make interoperable, data collected from a variety of ecosystem observational networks. Although only some of the most recent efforts have been designed with the impacts of climate change in mind, the process for integrating networks shared some common themes:

- Networks were often designed for a specific purpose, and are later pressed into service for new or emerging purposes;

- Data management and data interoperability are omnipresent challenges;

- Support and buy-in from data collectors are critical in determining continued support/longevity; and

- Integration efforts can take the form of "building from the ground up" new observational platforms or combining output from existing networks.

Following the workshop, the EIWG has attempted to synthesize many of these key issues into a framework for understanding the relationship between the inter-related goals of selecting indicators for the NCA and integrating observational networks. This framework (shown in Figure 2) presents a set of issues and questions that should form the basis of efforts to accomplish both of these goals. Specifically, for selecting indicators for the NCA process:

- The **lexicon** associated with the indicators should be established. From the workshop discussions, it is clear that many of the terms associated with ecology, climatology, and policy can be interpreted differently by those from different technical backgrounds. Examples include "indicator," "climate change," "monitoring," and "trend."

- The **scope and purpose** for a set of indicators should be determined. Choosing the types of decisions to which the indicators will be tied, the types of information the indicators need to convey, and the spatial and temporal scales at which the information should be provided all influence the process of indicator selection.

- The **audience** for a set of indicators should be identified. Closely tied to the issue of scope and purpose, it should be determined which group(s) will use the indicators. If multiple audiences are desired, then multiple sets of tailored indicators may be necessary.

Similarly, the framework identified sets of questions that would need to be answered as part of the process for integrating ecosystem observational networks.

- **Data collector/data user requirements**: How will the individuals and organizations that currently collect and use data from ecological observations be involved in the design of the networks? How and to what extent can their needs and perspectives be addressed?

- **Coordination requirements**: Which networks should be included, and who can represent these networks? How can buy-in be achieved among the parties involved?

- **Data requirements**: What types of data are available? To what extent can the data inform us about the impacts of climate change? What are the spatial and temporal aspects of the data? Do metadata exist that describe instrumentation or measurement protocols?

The issues and questions that are included in the framework are highly inter-related and it will be necessary to consider the features of the existing observational network when selecting indicators. Conversely, efforts to integrate observational networks could be heavily influenced by the set of indicators selected. For both the indicator selection and network integration efforts to be successful, it is likely that a high degree of coordination will be necessary in resolving the issues and questions outlined in the framework.

In addition to providing a summary and explanation of the key issues raised during the workshop and detailing the framework developed by the EIWG, this report concludes with some suggested next steps. These steps constitute significant commitments from a set of "experts" that would be tasked with addressing many of the issues within the framework, and a set of "envoys" that would be tasked with making connections among researchers, representatives from non-governmental organizations, and staff at federal, state, and tribal agencies with in-depth knowledge of the nation's ecological monitoring assets. A timeline outlining the activities of these two groups is presented, highlighting the important milestones to be pursued in 2011 and 2012.

INTRODUCTION

On November 30 and December 1, 2010, nearly 60 representatives from federal agencies, non-governmental organizations (NGOs), and scientific research institutions convened to discuss the observational data, monitoring networks, and indicators of climate change impacts on ecosystems in the United States.

Workshop Goals and Format

The Ecosystems Interagency Working Group (EIWG) hosted the workshop, which was sponsored by the Environmental Protection Agency (EPA). The EIWG meets under the auspices of the U.S. Global Change Research Program, and includes members from the Departments of Agriculture, Defense, Energy, and Interior, EPA, National Aeronautics and Space Administration, National Oceanic and Atmospheric Administration, National Science Foundation, and the Smithsonian Institute.

There were two goals for the workshop:

- **Contribute to the National Climate Assessment (NCA) process and products by identifying a set of indicators that represent the impacts of climate change on ecosystems**. It was hoped that the workshop attendees could help outline a process for selecting such indicators (e.g., who would be involved in the selection process, what criteria would be applied when selecting indicators), as opposed to generating a list of indicators. The ecosystem indicators, along with similar indicators representing the impacts of climate change across physical systems (e.g., impacts on the cryosphere or carbon cycle) and socioeconomic systems (e.g., impacts on public health or human settlements) will contribute to the NCA. The indicators will serve as part of the mechanism to make the assessment of climate change impacts a continuous, real-time process, rather than a series of episodic reports.

- **Support the EIWG's efforts to coordinate among existing observational networks to improve our understanding of the impacts of global change on ecosystems**. One of the EIWG's long-term goals is to determine the capability of existing monitoring and long-term observation networks to detect changes in ecosystem structure, ecosystem function, and biodiversity due to global change stressors. "Global change" includes climate change, large-scale land-use patterns, and changes in biogeochemical cycles. In working toward this goal, the EIWG is attempting to identify the observational networks available (or networks soon-to-be available); the types of data that are collected; the spatial and temporal characteristics of those data; the opportunities and barriers to synthesize data across different networks into useful information for researchers and decision makers; and the gaps in existing networks and future observational needs. Through the workshop, the EIWG hoped to gather more information about observational networks, and more importantly, initiate a dialog among the individuals and organizations that could contribute to these efforts in the future.

These two goals are highly related. In Section IV, a framework for pursuing these two goals is laid out, based on the input of workshop participants and subsequent discussions of the EIWG.

Presentations were made regarding past efforts to develop sets of ecosystem indicators and to combine or integrate data from across a variety of observational networks. Many of these presentations are available at https://sites.google.com/a/usgcrp.gov/eco-monitoring/. Workshop attendees participated in one of three ecosystem-specific breakout groups: terrestrial, freshwater, and marine. The breakout discussions explored how indicators might be selected for the respective ecosystem types, as well as what types of observations are available (or might be made available) for each type of ecosystem. In a plenary session, the workshop attendees discussed the needs and vision for the NCA.

Report Goals and Format

This report is intended to summarize the discussions that occurred at the workshop; to communicate the post-workshop discussions of the EIWG regarding the selection of indicators and the integration of observational networks; to outline a framework for making progress toward the goals of the NCA and the EIWG; and to suggest several next steps for achieving these goals.

The remainder of this report is organized into three sections:

- Section III - A distillation of key points raised in the workshop presentations, discussions, and background materials;

- Section IV - The introduction and explanation of a framework to make progress toward the dual goals of: 1) contributing to the selection of a set of ecological indicators for the NCA, and 2) fostering collaboration among federal agencies that collect and manage data on ecosystem structure, ecosystem function, and biodiversity that could be used to understand the impacts of global change across the nation; and

- Section V - A time line and set of next steps that the can be pursued to achieve these two goals.

Summaries from the Breakout Groups

Participants split into three breakout groups, each focusing on a subset of ecosystem domains: marine ecosystems, terrestrial ecosystems, and freshwater ecosystems. Each group's discussions were distinct – although similar themes emerged among groups (and form the basis of the Key Issues raised in Section III), the trajectory and topics of their respective discussions were unique. The salient points from each group are summarized below.

The **Marine Breakout Group** discussed criteria for selecting indicators, focusing on:
- Aspects of data collection, including the current availability of data (*i.e.*, are the data being collected now) and the spatial and temporal coverage of data; and
- The indicator's sensitivity to climate change.

Several types of indicators were suggested, including phenology, species abundance, distribution, and diversity (including information on invasive species), and primary productivity. Through an informal voting process, these indicators were scored according to the two criteria, how well or poorly the indicator is observed and how sensitive it is to climate change. Phenology scored "high" (*i.e.*, it was considered to be relatively well observed and highly sensitive to climate change), while invasive species scored "low" (*i.e.*, poorly observed and their sensitivity to climate change is incompletely understood).

The **Terrestrial Breakout Group** raised some key questions regarding the purpose of selecting a set of indicators and the elements of a selection process, including:

- Who is the audience for the indicators?
- How will the indicators be used?
- Who should be involved in the selection process?
- How will coordination of the selection process among national to regional scales occur?

The group discussed the types of information an indicator should represent (i.e., whether an indicator should capture the status and relatively slow, structural changes in a system, or whether it should provide "early warning" regarding tipping points or thresholds). Also, the group raised questions regarding the adequacy of monitoring networks for providing such information at the appropriate spatial and temporal scales. The group also noted the need for a multiple stressor approach. Difficulty understanding and detecting the impacts of climate change requires an understanding and observational capacity across all relevant environmental stressors.

The group discussed two potential indicators: nutrient ratios/concentration (specifically nitrogen) and species distribution. These indicators were chosen since they could be considered "leading indicators" – changes in these indicators might be detected prior to or in anticipation of more fundamental changes or shifts in the structure and function of an ecosystem.

The Freshwater Breakout Group discussed the purpose of indicators as being "integrative" – condensing complex information about the state, flows, and processes of an ecosystem into simpler terms that are useful for understanding climate change in both decision-making and research contexts. Participants noted the need for both biotic and abiotic information when establishing the case for attribution of observed changes. They also discussed some of the important data considerations (e.g., availability, transparency, length of record, documentation of metadata) when choosing indicators.

The group created a list of aspects of useful indicators:

- Viewed as legitimate by stakeholders;
- Connected to vulnerability assessment and management decisions;
- Part of a model or narrative that clearly links climate change to ecosystems;
- Effective as communication tools; and
- Contain early warning information about tipping points.

The group expanded on the desire for early warning information and compiled a short list of potential indicators, all of which were considered to be highly sensitive to changes in climate, especially temperature.

The group also noted the needs for a data management structure and an institutional infrastructure (*i.e.*, individuals and groups that are committed to the collection of data for indicators, and the deployment of indicators in decision-making) to ensure that the efforts to maintain and use a set of indicators would be sustainable.

BACKGROUND INFORMATION/WORKSHOP SUMMARY

The workshop participants discussed past efforts to combine or integrate ecosystem observational networks as well as efforts to develop systems of ecosystem indicators. These discussions compared and contrasted some of the past activities and highlighted lessons learned from their successes and failures.

Indicators

In discussing past work[1], the workshop attendees noted several distinguishing features of indicators:

- Indicators simplify and condense complex information. This can serve as a crucial step toward communicating information about ecosystems.

- Indicators are measureable and quantitative. Although some previous work employs qualitative indicators, most of the workshop discussion treated indicators in a quantitative sense.

- Indicators capture information about changes over time, such as trends or changes in variability. Indicators are often portrayed as a time series of values.

Some examples of potential indicators and their relationship to environmental measurements can be found in the NRC report Monitoring Climate Change Impacts: Metrics at the Intersection of the Human and Earth Systems (National Research Council (NRC) 2010; see Table 1). Although the NRC indicators were crafted to address overall environmental sustainability and its relationship to human systems, the explanations associated with these indicators remain useful and illustrative for discussing the impacts of climate change.

A series of key issues regarding indicators raised during the workshop presentations, plenary discussions, and breakout groups are described below.

---

[1]Appendix A and Appendix B provide information from past work about sets of indicators and the basis for their selection, respectively. Rather than systematically analyze these lists, the workshop participants focused primarily on deeper issues associated with the purpose and design of a set of indicators. These deeper issues are discussed in this section and the following section. The Appendices are provided as illustrative references - they are not intended to advocate for the adoption of a particular set of indicators or selection process.

Indicator	Measurements	Link to Environmental Sustainability
Sea level rise	Global sea level height Glacial (ice) measurements High-resolution maps of terrestrial features Advanced circulation models of inundation Sea floor depth	Temporal and spatial patterns of changes in sea level will be an indicator of future risks to coastal populations and infrastructure. Higher sea level amplifies coastal erosion, storm damage, permanent flooding, and land inundation.
Soil moisture change	Soil moisture Plant productivity Decomposition rate Soil formation rate	Soil moisture is a major controlling variable for large-scale patterns in vegetation. Soil moisture dynamics are critical variables for many of the ecological models used for global carbon budgets and other global ecological processes.
Mass of small, high altitude glaciers	Glacier extent in summer and winter Surface elevation in summer and winter	Loss of these glaciers would remove a critical water source (especially in summer) for many high-elevation populations.

Table 1: Indicators, their relationships to measurements, and their links to environmental sustainability. SOURCE: Adapted from National Research Council (NRC), 2010.

Key Issue: Determining the Purpose of a Set of Indicators

Sets of indicators exist for a wide range of purposes (see Appendix A, which includes several example sets of indicators). For example, in Ecological Indicators for the Nation (National Research Council (NRC), 2000) and in reports from the Heinz Center (Heinz Center, 2002, 2008), the indicators were designed to summarize the health of ecosystems on a broad, national scale. The indicators focused on measures of the state, functionality, and productivity of ecosystems (*e.g.*, land use, species diversity, and nutrient-use efficiency) that could be aggregated across a wide range of diverse ecosystems. These indicators were intended to inform national-level policy decisions on such ubiquitous issues as land use change, water quality, habitat protection, and the regulation of pollution. However, such indicators are not always applicable to more local-scale resource management decisions. Similarly, these types of indicators are not designed to establish attribution between observed changes in ecosystems and large-scale stressors (*e.g.*, climate change, ocean acidification, degraded air quality).

The limitations are not noted as a criticism of these sets of indicators, or the process of their selection[2]. Rather, they demonstrate that the intended purpose for a set of indicators is closely tied to 1) the criteria applied within the selection process, and 2) the ultimate efficacy of applying the indicators to relevant scientific, policy, or management issues.

One particularly important aspect of an indicator's purpose with regard to climate change involves its predictive capability. Indicators regarding ecosystems would seem to have two distinct and somewhat exclusive purposes: they could be designed to be diagnostic, informing about changes occurring or that have occurred within an ecosystem; or, they could be prognostic and provide an "early warning," notifying or predicting an impending (likely negative or undesirable) change within the ecosystem. Not only could these two purposes lead to the selection of different sets of indicators, but the two purposes introduce different biases. With regard to climate change, the early warning indicators are likely to be highly sensitive to some subset of cli-

---

[2]To the contrary, the efforts by the NRC and the Heinz Center are notable within the ecological community for their careful documentation of an appropriate selection process.

mate variables, such as temperature or precipitation. Accordingly, they may be considered biased in that they will exhibit greater change (in a normalized sense) than other indicators or measurements when perturbed by a specified change in temperature or precipitation. More diagnostic indicators, on the other hand, may be biased in a conservative sense if they focus on the slowly changing parts of an ecosystem (e.g., sea level, soil carbon). Although such diagnostic indicators may be well correlated to other components of an ecosystem, these slowly responding indicators are likely to aggregate changes that have already taken place among the processes with faster timescales. Essentially, they will record changes or trends after they have taken hold in an ecosystem.

Key Issue: Using Indicators to Understand Attribution or Cause-and-Effect Relationships

Only the most recent work on ecological indicators has had an explicit focus on the impacts of climate change (e.g., Environmental Protection Agency (EPA), 2010; National Research Council (NRC), 2010); much of the earlier and more in-depth work (e.g., Heinz Center, 2002, 2008; National Research Council (NRC), 2000) focused on the overall state and function of ecosystems as they are confronted with a variety of stressors.

The extent to which indicators can be used to attribute impacts to climate change was a point of contention among workshop participants. Although there is a desire to use indicators to summarize and communicate the effects of climate change on ecosystems in a particular region or for the country as a whole, doing so requires observation and understanding of other non-climate drivers (e.g., land use change, air or water pollution) that might also be affecting these ecosystems. Interactions between climate change and non-climate drivers could be additive, offsetting, or multiplicative (or more generally, non-linear). Efforts to disentangle the influence of multiple stressors are likely to require consideration of biotic information (e.g., biodiversity, phenology) in conjunction with abiotic information (e.g., soil moisture, nutrient levels and nutrient-use efficiency), as well as indicators for the physical climate system (e.g., temperature, precipitation).

The need for cause-and-effect information not only has implications for the design of an indicator and its underlying measurement data, but also for the

indicator's role in decision-making. In some cases, linking an indicator to a particular management action or policy measure may not require information on attribution. For example, if an indicator suggests that a particularly important species is in decline within an ecosystem, action might be taken to protect or relocate that species, regardless of the precise factor(s) responsible for the decline. In other cases, attribution information will be integral to shaping a management or policy response, or in evaluating the impact of previous management practices. For example, understanding the relative roles of climate change versus local land use change in affecting an indicator (and thus the ecosystem as a whole) would be valuable in formulating and prioritizing adaptation options.

Key Issue: Tying Indicators to a Conceptual Model of Ecological Function and Ecosystem Services

From a scientific perspective, indicators represent a conceptual model that captures how changes in climate are expected to influence the processes that occur within ecosystems. These processes can span many spatial scales, from the landscape level to the regional or continental level. They can also involve coupling across these scales, as changes at smaller scales can act to influence an ecosystem on a larger scale, and vice versa. Likewise, the influence of climate change on ecosystems can span many timescales, including responses to episodic events (e.g., disturbances such as floods or forest fires), slower changes in the timing and characteristics of ecosystems over years and decades (e.g., the migration or fragmentation of a forest habitat), and the coupling or feedbacks that arise between changes occurring at the various timescales.

It was beyond the scope of the workshop to define all the aspects of a conceptual model that govern how indicators represent the state of ecosystems or the key processes operating within an ecosystem. Example conceptual models exist that show the connections of some of the key climate drivers and ecosystem responses (Figure 1). Although this schematic is not comprehensive or quantitative, it captures some of the important aspects of climate change and disturbance that can influence ecosystems, the feedbacks involved between ecosystems and the climate system, as well as the link between the state of an ecosystem and its ability to render goods and services.

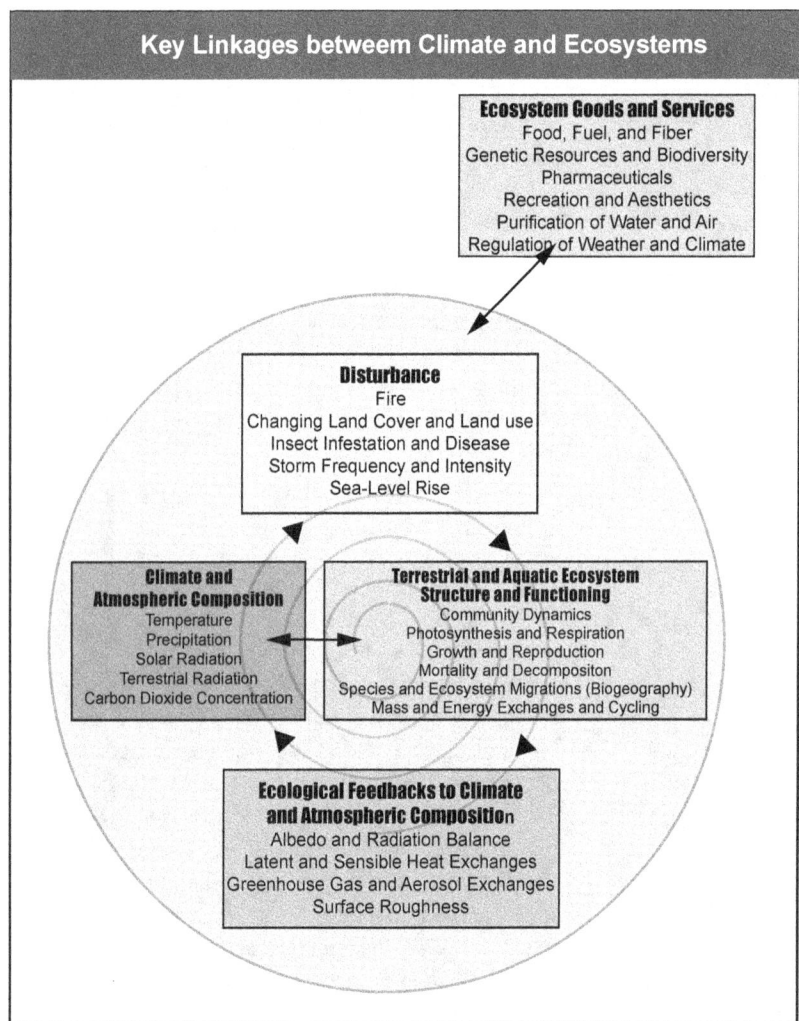

Key Linkages betweem Climate and Ecosystems

Ecosystem Goods and Services
Food, Fuel, and Fiber
Genetic Resources and Biodiversity
Pharmaceuticals
Recreation and Aesthetics
Purification of Water and Air
Regulation of Weather and Climate

Disturbance
Fire
Changing Land Cover and Land use
Insect Infestation and Disease
Storm Frequency and Intensity
Sea-Level Rise

Climate and Atmospheric Composition
Temperature
Precipitation
Solar Radiation
Terrestrial Radiation
Carbon Dioxide Concentration

Terrestrial and Aquatic Ecosystem Structure and Functioning
Community Dynamics
Photosynthesis and Respiration
Growth and Reproduction
Mortality and Decompositon
Species and Ecosystem Migrations (Biogeography)
Mass and Energy Exchanges and Cycling

Ecological Feedbacks to Climate and Atmospheric Composition
Albedo and Radiation Balance
Latent and Sensible Heat Exchanges
Greenhouse Gas and Aerosol Exchanges
Surface Roughness

Figure 1: A schematic representing the interactions between climate and ecosystems. Climate can influence ecosystems directly (represented by the arrows connecting the "Climate" and "Ecosystem" boxes). Indirect impacts that arise from disturbances or feedback mechanisms are also represented (spiraled arrows). These interactions can impact the goods and services provided by ecosystems (upper right box). The complexity and sometimes non-linear character of these interactions can make it difficult to assign cause-and-effect relationships for observed changes in ecosystems.
SOURCE: Climate Change Science Program (CCSP), 2008

cover type), capital (*e.g.*, species diversity, soil organic matter), and function (*e.g.*, net primary productivity, nutrient-use efficiency).

Identifying feedbacks or "tipping points" within an ecosystem represents a challenge when selecting indicators. The strength of a feedback or the presence of a "tipping point" may not be known currently – using indicators to monitor such behavior, or more generally, establishing how to use indicators to monitor such behavior, may require more research.

During the workshop discussions, two ideas emerged for indicators that specifically focus on climate change: monitoring changes in the phenology of plants and animals (*e.g.*, bloom dates or emergence dates) and tracking changes occurring along ecological gradients (often across altitude or latitude). These ideas for indicators (see EPA (2010) for further discussion of these types of indicators) demonstrate that simply adopting the conceptual basis associated with previous efforts to select indicators may not always lead to a comprehensive treatment of the impacts of climate change. Essentially, our expectations of how ecosystems might change in response to changes in climate should be codified in any set of indicators that is selected.

In addition to the scientific and technical issues associated with selecting a set of ecological indicators, the workshop participants raised some important programmatic and organizational issues. These are described as the next key issues.

Other aspects of a conceptual model discussed at the workshop that might underlie a set of indicators involved the monitoring of the area occupied by an ecosystem, flows of energy and nutrients into and out of an ecosystem, and the types and abundance of species found in an ecosystem. Many of these ideas are generally consistent with explanations found in Ecological Indicators for the Nation (National Research Council (NRC), 2000), in which the indicators of ecological status (which were chosen broadly, without a particular emphasis on climate change) focus on ecosystem extent (*e.g.*, land

Key Issue: Designating the Parties Involved with the Selection Process

Among the previous studies, and as discussed at the Workshop, there are a variety of processes that have been pursued in selecting a list of indicators. For the NRC reports (National Research Council (NRC), 2000, 2010), a set of academic experts representing a range of disciplines was assembled. Although these reports underwent rigorous peer review, the lists are essentially a compilation and consensus of expert scientific opinions. The selection process for Heinz indicators was similar, but involved a wider range of parties from academia; federal, state, and local government; environmental non-governmental organizations; and the private sector. In addition to scientific experts, the Heinz process involved experts from legal, economic, social, and policy backgrounds.

The selection of indicators for the NCA is likely to undergo a process similar to the NRC and Heinz efforts. Input from a variety of scientific experts and a broad set of decision makers or other assessment users will be included. The NCA indicators will also be subject to a peer review process.

Key Issue: Achieving Buy-in from Users and Establishing an Association with Management Decisions

The extent to which managers can incorporate indicators into their management decisions has a sizeable influence on the indicators' adoption and longevity. The inability of managers to understand or utilize indicators was cited as a reason why several past efforts to generate a list of indicators have had little impact beyond academic or policy circles.

OBSERVATIONAL NETWORKS

Within the workshop, the discussions of indicators and observational networks were not formally segregated and overlapped considerably. Many of the above points regarding indicators could be applied to the design of observational networks, or to the attempts to use existing observational networks to glean information about the impacts of climate change on ecosystems. However there were several **key issues especially relevant to observational networks** that emerged from existing literature (see Heinz Center (2006) in particular), past experience highlighted by workshop speakers, and comments made by workshop participants.

Key Issue: Networks Designed for Specific Purposes Have Varying Abilities to Evolve to Meet New or Emerging Information Needs

Learning about the impacts of climate change on ecosystems can be facilitated by long time series of data (typically multiple decades of data), where information regarding instrumentation, measurement methodologies, and data format are all available. These data sets have the critical "ingredients" that allow them to be used in multiple contexts. Such data sets, and the networks that record such data, are likely "adaptable" to answering questions about climate impacts on ecosystems, despite the fact that they may not have been designed to do so.

Unfortunately, a number of still-operating networks that potentially record information related to climate change do not meet these "adaptability" criteria. Few of the networks deployed prior to the last decade had the mission of observing impacts of climate change. These older networks are likely to have been deployed with a local or regional focus, and it may be difficult to extrapolate their data to sufficiently large spatial scales to diagnose the influence of climate change (*i.e.*, it is difficult to separate the impact of local-scale ecosystem drivers, like land use change, from the impact of a large-scale forcing, like climate change). Alternatively, difficulties with instrument calibration and data validation are common among these relatively long-lived networks. Repairs and replacement of instrumentation or changes in sampling locations or frequency can make it difficult to interpret trends (or other features) in the time series of their observations.

Key Issue: Data Management and Data Interoperability are Omnipresent Challenges

Issues of data formatting, data access, data quality, continuity of observation, agency jurisdiction, and availability of metadata can inhibit the synthesis of similar data sets across disparate networks. This can occur even across networks that sample the same or similar variables.

As an example, workshop participants discussed some of the difficulties and challenges in identifying watersheds with long-term records of streamflow and water quality that could be used to asses impacts of climate change. Although many agencies have observed these variables over the past several decades, it is not a trivial task to identify sites where long-term records exist, observations are continuing, and data are of sufficient and consistent quality. Integrating the datasets to create a holistic, regional, or national understanding of changes in streamflow and water quality over time has been a substantial challenge.

Key Issue: Support and Buy-in from Data Collectors are Critical In Determining Continued Support/Longevity

For many observational networks, the managers "on the ground" will ultimately be responsible for installing and maintaining observational equipment. Although they can be provided with resources to carry out these activities, that alone is not always a sufficient incentive. Looking forward, it may be more difficult to garner such resources. However, if the observations can contribute to relevant management decisions and if managers can participate in the design and implementation of the overall program, the network is more likely to continue for a long time. Also, this buy-in and participation by managers ensures that an observational network will be flexible – the network can be altered over time to be responsive to the evolving needs of managers.

Participants also attempted to answer, "What makes an observational network successful at achieving and maintaining such buy-in?" Aspects of the National Atmospheric Deposition Program (NADP) were mentioned during this discussion since the NADP has created a process whereby members of federal and state agencies have partnered with NGOs and private, academic, and tribal groups to make decisions about the observational networks associated with the program. These groups have used the data for their management and research purposes, and their feedback has contributed to the program's evolution over time.

Key Issue: Building from Ground Up vs. Integrating Existing Networks

Two current efforts, the National Ecological Observatory Network (NEON, http://www.neoninc.org/) and the Climate Effects Network (CEN, http://gcp.usgs.gov/cen/) are attempting to deploy/integrate ecological observations that can inform researchers and resource managers about the impacts of climate change.

Although their goals are similar, the strategies of NEON and CEN are quite different. The NEON program is establishing new observational sites that are coordinated with laboratory experiments, remote sensing platforms, and model develop-

ment. All activities are integrated using common standards for data management. The systems have been designed to continue for at least 30 years, providing long-term assessments of impacts from environmental stressors (e.g., climate change, land use change) at the regional to continental scales. On the other hand, the CEN is attempting to utilize a wide range of existing observational networks (e.g., National Water-Quality Assessment Program, Forest Inventory and Analysis National Program, National Phenology Network). Its activities are intended to be guided both by a need to inform local resource management decisions and to lay the groundwork for scientific investigations of the impacts of climate change. A CEN pilot program for the Yukon River Basin has begun, and programs regarding biological carbon sequestration and a national water census are in the planning phases.

A FRAMEWORK FOR SELECTING A SET OF INDICATORS AND INTEGRATING ECOSYSTEM OBSERVATIONAL NETWORKS

Following the workshop, the EIWG recognized and acknowledged its role in pursuing two interconnected goals, displayed visually in Figure 2:

- To contribute to the NCA's efforts to develop a set of indicators that represent the impacts of climate change on the nation's ecosystems (top, red box). These indicators would be a subset of a larger collection of indicators that would represent impacts across the physical climate system and socioeconomic systems. These indicators would serve as part of the mechanism to make the assessment of climate change impacts a continuous, real-time process, rather than a series of episodic reports.

- To determine the capability of existing monitoring and long-term observational networks to detect changes in ecosystem structure, ecosystem function, and biodiversity that are related to global change stressors (bottom, blue circles). "Global change" includes climate change, large-scale land use patterns, and changes in biogeochemical cycles. As part of this analysis, the ability and barriers to synthesize data from different networks into useful information for researchers and decision makers

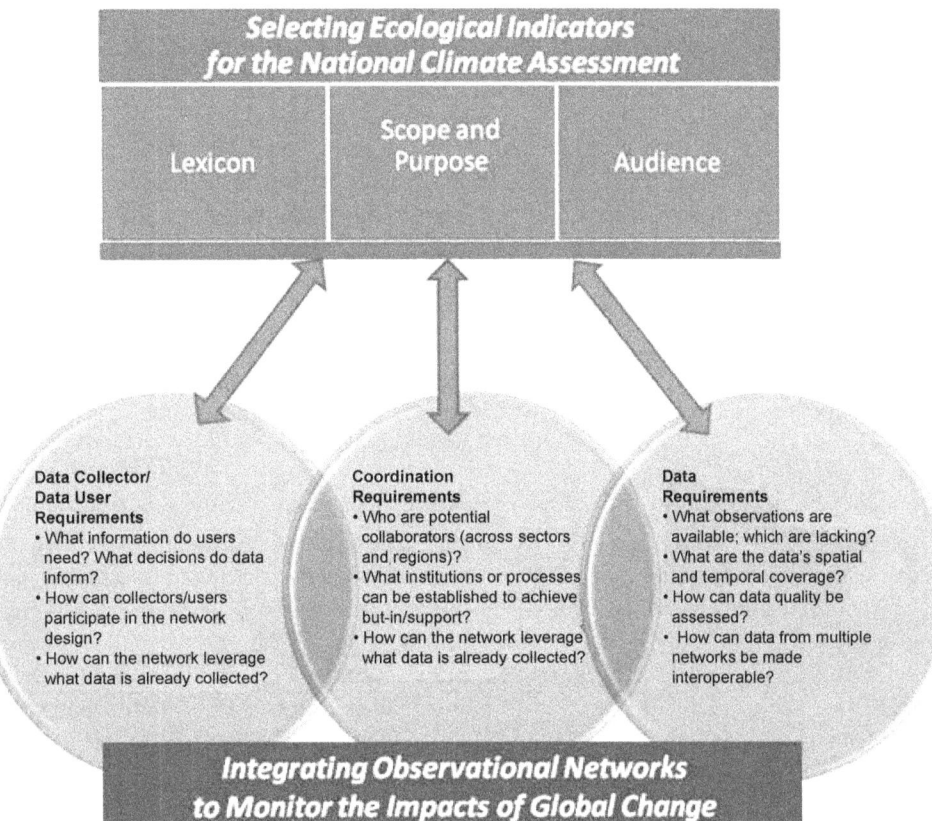

Figure 2: The framework envisioned for selecting indicators for the National Climate Assessment (top of diagram, red boxes) AND for integrating observational networks (bottom of diagram, blue circles). Each box/circle contains a key issue or question that must be addressed in order for the EIWG to make progress toward these two goals. The diagram also emphasizes the connections between these two goals.

would be considered. Gaps in existing observational networks and future needs would be identified. The EIWG would aim to enhance cooperation and coordination across federal agencies engaged in ecological monitoring and to leverage the efforts across these agencies.

The visual depiction of the framework is intended to lay out these two goals, each with a requisite set of questions or issues that are to be resolved. Although many of these issues were raised at the workshop, the format and time allotted to the discussion did not permit the group to engage in a comprehensive deliberation leading to any sort of resolution. The Next Steps section outlines a series of potential activities that might lead to some resolution of these issues.

These questions and issues provide structure to the EIWG's pursuit of the two goals. As shown in Figure 2, the needs and efforts associated with the

selection of indicators can be thought of as "top-down." The NCA process is driven by the need for a comprehensive national-level summary of the impacts of climate change across various regions and sectors. It synthesizes the results and experiences from many federal agencies, as well as from many experts operating at state and local levels. In contrast, the integration of observational networks can be thought of as a "bottom-up" exercise.[3] It is driven by the types of data collection activities that are occurring or could occur in terrestrial, freshwater, and marine ecosystems, as well as across spatial scales (e.g., by satellites or other remote sensing platforms). By assessing these disparate networks, as well as the existing efforts to coordinate among them, the EIWG can identify areas of redundancy

_____

[3]As mentioned in the Key Issues section, the success of the observational integration will require processes that are designed to include and achieve buy-in from many federal agency staff as well as tribal, state, and local decision makers. Unless these parties are vested in the process and can use the data products from the network to make progress toward their respective management goals, then the motivation to integrate will likely be lacking.

and opportunities for collaboration and coordination across federal agency efforts. This can aid federal agencies as they set priorities for ecological monitoring, particularly to detect the effects of climate change. If well executed, such an assessment could generate information regarding climate change effects that would be useful for resource managers working at a variety of spatial scales, from local to national.

Going Forward: Selecting Indicators

The three categories of issues to be resolved as a prerequisite for selecting a set of indicators are shown in the red boxes. These issues are themselves connected to one another. For example, definitions for key terms (lexicon) will be linked to the scope and purpose of the indicators, as well as to the identities of the intended audiences for the indicators. Below, each of these indicator-related issues is described in more detail.

Clarifying the Lexicon

Technical terms associated with climate science and climate policy have a variety of interpretations and attributes, depending upon the disciplinary context of their use (see "Terminology" Box below). The imprecision likely reflects the relative "youth" of the climate science and climate policy fields when compared to more established fields (e.g., biology, chemistry), as well as their inherent interdisciplinarity.

The process of selecting indicators will require extensive, ongoing discussions with a variety of experts including scientific researchers, resource managers, policy analysts, and federal, state, and local government representatives. Effective dialog among these groups will require clarity, which requires the establishment of "ground rules" for the usage of many terms. For example, unless specified, even the simple phrase "climate change" can be problematic. For example, is climate change restricted to the statistics of weather, or should it include biogeochemical changes as well? And does climate change only refer to anthropogenically-forced changes in such variables, or any changes, whether natural or anthropogenic, in such variables? Resolving linguistic ambiguity will also be crucial for communication purposes and adoption of the indicators by those researchers, agency staff, resource managers, and decision makers not involved in the process of their selection.

It should be noted that the establishment of the lexicon is not an academic exercise (i.e., it is not being done to suggest some optimal set of definitions to be adopted), but is specifically linked to the process

---

Terminology

Differences in the uses and meanings of certain terms acted as a stumbling block during several workshop discussions. The attendees hailed from a wide range of academic and professional disciplines, and their usage of key terms reflected the subtle differences in connotation that can be found across these disparate disciplines. Some examples include:

- What exactly is an *indicator*? How does it relate to *measurements, observations,* or *monitoring*? Are these three terms interchangeable, or are there distinctions among them?
- Are there substantive differences when discussing the *consolidation, integration,* or *interoperability* of data?
- What is meant when one refers to a *user* or a *decision-maker*?
- Is the term *climate change* limited to changes in the statistics of weather, or does it extend to other changes in the earth system (e.g., ocean acidification)? Is it limited to anthropogenic climate change, or is it inclusive of both natural and anthropogenic changes?
- To what does a *trend* refer? Is it restricted to linear changes over time, or does it broadly represent many types of changes (e.g., regime shifts, changes in variance) that might be extracted by analyzing a time series of data?

Although the workshop participants did not have sufficient time to determine the best ways to define these terms, it is clear that some clarification is necessary when engaging in interdisciplinary discussions regarding ecosystem indicators or ecosystem observational networks.

of selecting indicators. It is a means of establishing a clear, common, and agreed-upon set of terms that codifies the purpose of the indicator-selection effort; its establishment is meant to prevent the constant rehashing of semantic arguments among individuals that will be expected to contribute to the process or utilize the output of the process.

Establishing the Scope and Purpose of the Indicators

A prerequisite to forming a list of indicators is the establishment of their scope and purpose. Some questions to be answered include:

- Should indicators provide "early warning" to resource managers or policy makers to take action? Or should they be diagnostic, reflecting changes as or after they occur in ecosystems? These differing goals would likely lead to different choices for indicators. For example, for a terrestrial early warning system, indicators could focus on microbial communities with short lifetimes and/or high sensitivity to climate conditions; for a diagnostic system, indicators would likely focus on slower changing portions of an ecosystem, like the amount of soil carbon present. As mentioned in Section 3, these types of choices will also involve some degree of bias in how the indicator portrays the pace and magnitude of the impacts of climate change.

- What types of decisions will be made using indicators? Will they involve setting priorities for science research? Or involve identification of climate adaptation options or other resource management options? Or, rather than being designed for decision-making, are the indicators primarily tools for communication? If multiple purposes are desired, what is the process for selecting the different subsets of indicators?

- Are there certain ecosystems (or species) that should receive a high priority, perhaps based on their sensitivity or vulnerability to climate change (e.g., some Arctic ecosystems) or their provision of ecosystem services (e.g., commercial forests, agriculturally important ecosystems)? If ecosystem services are important, how should they be evaluated, given that there are a variety of economic and non-economic metrics

available? If multiple purposes are desired, what is the process for selecting the different subsets of indicators?

Answers to these questions will determine the spatial and temporal scales upon which the indicators should focus (e.g., indicators tied to a threatened species found in the Everglades would be quite different from indicators capturing large-scale decadal changes to forests across the western United States). The manner in which the indicators address attribution of change is also connected to these answers. If assessing vulnerability and/or providing early warning are considered part of the indictors' purposes, then attributing ecosystem changes to climate change (as opposed to other environmental stressors) may not be critical. In these cases, knowing that an ecosystem has experienced some level of degradation or is highly vulnerable to future, near-term degradation may be a sufficient basis for action, regardless of the precise cause of the past or future degradation. However, for larger-scale assessments or more academic or scientific endeavors, there may be a premium on attributing observed or expected changes specifically to climate change, or to enhancing our overall understanding of the complex relationships and responses that emerge within ecosystems coping with multiple stressors.

From a more pragmatic perspective, these questions also highlight which parties should be involved with the process of selecting indicators and what types of data will be necessary to generate a time series of indicators. With regard to the latter, the indicators may require information about changes in the physical climate system and socioeconomic systems, in addition to information from ecosystems. Also, whether the indicators should be drawn from existing data, or whether they could be selected in the hopes of developing new observations should also be determined as a matter of resolving the indicators' scope and purpose.

Once the scope and purpose for the indicators have been settled, the process of deciding what makes a "good" or "bad" indicator, and the criteria for doing so, can be explored. The NRC 2000 report lists such a set of criteria (see Appendix B). Although the NRC listing is considered to be a benchmark for other efforts to compile lists of indicators, the report's purpose was explicit: it was designed to identify "national-level indicators to inform major policy decisions" and its indicators are not necessarily designed to be linked to climate change, to be used as communication tools, or to be a tool for local or regional management (although they could potentially be adapted to serve these other purposes).

Establishing the Audience(s) for the Indicators

Determining the audience(s) for the NCA indicators is another important prerequisite for their selection. It is likely that multiple audiences will be targeted, including policy makers, resource managers, and scientific researchers. These different groups might be concerned with issues on national, state, or local levels. Consequently, effectively addressing these disparate audiences may require different sets of indicators, each having been designed with that audience's goals in mind.

Figure 3 categorizes some of the participant perspectives from the workshop to illustrate their respective and somewhat distinct near-term and long-term goals. Although most of the individuals in attendance would likely consider themselves part of more than one of these groups, it was clear from the discussions that subtle differences among the attendees' goals can lead to different values and prioritizations that affect their potential need for and use of indicators. Thus, the targeted audience(s) for any set of indicators should be identified – both as part of the process of selecting the indicators and in ensuring that their deployment leads to their actual use.

Going Forward: Integrating Observational Networks

Several inter-related challenges (blue circles, Figure 2) underlie an effort to integrate existing observational networks; these are outlined below. Although the goals associated with this integration are slightly different than those associated with selecting indicators, addressing the three challenges

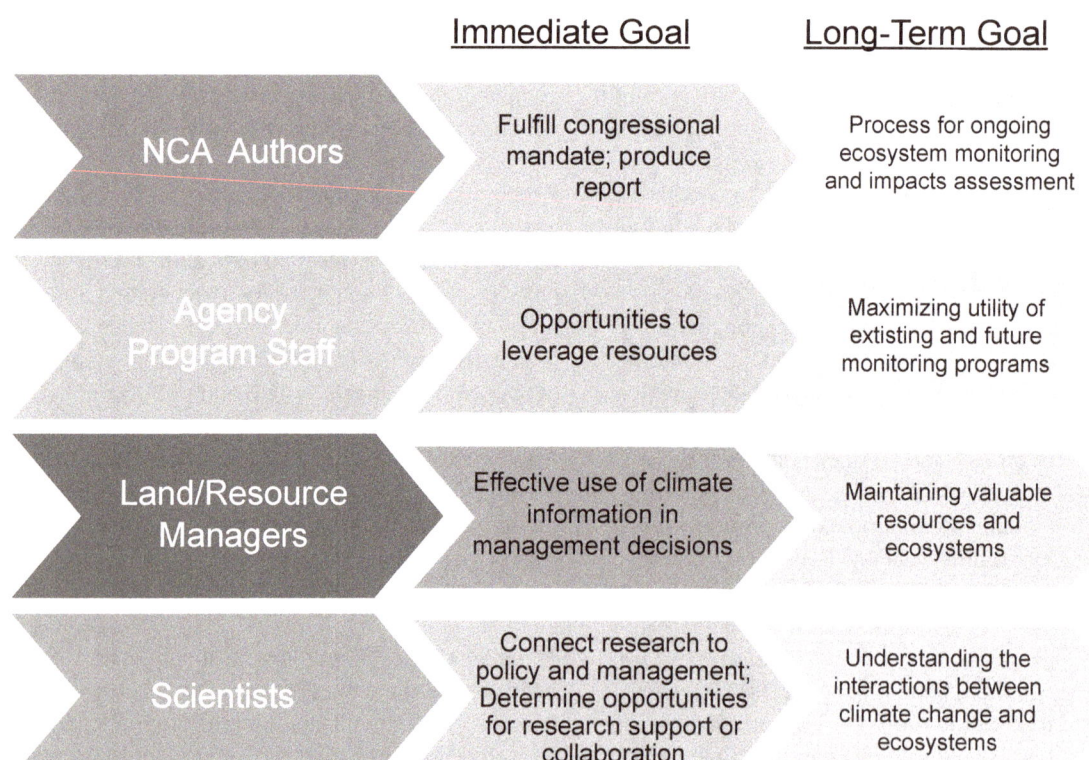

Contributors: Goals and Perspectives

	Immediate Goal	Long-Term Goal
NCA Authors	Fulfill congressional mandate; produce report	Process for ongoing ecosystem monitoring and impacts assessment
Agency Program Staff	Opportunities to leverage resources	Maximizing utility of extisting and future monitoring programs
Land/Resource Managers	Effective use of climate information in management decisions	Maintaining valuable resources and ecosystems
Scientists	Connect research to policy and management; Determine opportunities for research support or collaboration	Understanding the interactions between climate change and ecosystems

Figure 3: Description of the goals and perspectives of workshop attendees. Although many of the workshop attendees focused on ecological issues at the national level, some were more focused on regional and local dimensions of the above goals. The diagram is not meant to be comprehensive, or exclusive (many individuals might fit into multiple categories), but to illustrate the subtle differences among the agendas of individuals or institutions involved in the processes of selecting ecological indicators and integrating observational networks.

(described below) is closely connected to resolving the issues outlined for the indicator selection. For example, without some sort of inventory of the types of observations and data available, it will be difficult to deploy or implement indicators in the near-term. Likewise, if the agencies and personnel responsible for the collection of data and maintenance of observational networks are not involved (or at least considered) in the process of selecting indicators, the continuance of the observations could be at risk.

Data Collector and Data User Requirements

As previously mentioned, buy-in from and participation of those individuals and groups that collect and use the data from observational networks have been key components in the success of past programs[4]. With regard to data collectors, understanding their operational needs and leveraging their ongoing data collection activities will be critical, as it is unlikely that they have the time or resources to commit to new initiatives for network integration that do not overlap with their own goals and missions. Similarly, it will also be important to consider the needs of individuals and groups that currently use observational data to make management decisions, enforce regulations, or perform scientific research. Consideration of the types of data that are useful for them and the types of new data that they would like to have will add legitimacy to the efforts to integrate observational networks and/or identify new high-priority observations.

Coordination Requirements

Two tasks are required to achieve effective coordination. First, potential collaborators must be identified. These individuals must have expert knowledge about the available observational networks, the maintenance of these networks, and the use of the data generated by these networks. These collaborators would likely hail from a range of federal, tribal, state, local, non-governmental, and/or academic organizations involved with scientific research, environmental policy, and land/resource management. Incentives for collaborators to participate

[4]The titles "Data Collector" and "Data User" may not be the best terms to use. As part of clarifying the lexicon associated with the process of selecting indicators, broad terms like these may need to be more precisely defined. In many cases, the collectors and users of data may refer to the same groups. Also, the term "user" is intended to be broad. It could include land and resource managers operating at the local level, or those using the output of observational networks for research activities, which might include modeling studies or impact assessments.

and maintain engagement over time should also be identified.

Second, processes or institutions must be identified that provide a forum for these collaborators to meet and discuss the capabilities and needs of ecosystem observational networks. Although some new institutions may need to be developed (e.g., a set of interagency workshops), existing or planned activities should be leveraged (e.g., occasions when these potential collaborators might already be meeting should be identified).

Data Requirements

Identifying impacts of climate change on ecosystems will likely require data from a diverse set of observational networks operating across multiple geographic regions. In order to make data interoperable across networks, aspects of data quality, accessibility, and transparency will need to be assessed. Standards may need to be developed, or adopted from current efforts, to link observational networks. Examples of such data requirements include:

- Spatial and temporal characteristics: Are measurements representative of local conditions only or would they represent similar ecosystems in other regions? How far back does the data record extend? Are measurements ongoing or only valid for certain periods of time? Are measurements made at regular intervals?

- Data formats: Are data saved in an electronic format, or some other medium? How are data time-stamped and geo-referenced? What computing, including hardware and software, or statistical resources are necessary to manipulate and visualize the data?

- Documentation and transparency: Access to metadata related to instrumentation, missing data, and calibration techniques, can be critical for interpreting data, or attempting to combine them with other observations. Can a user determine how the data were collected? To what extent have the data been quality controlled? Do the data contain any known biases or caveats?
- Data access and management: Can both federal and non-federal entities access the data? Are data available online? Which organizations are best suited to undertake such tasks? If lead organizations are desig-

nated, what process(es) exist for involving other organizations that produce and use the data? As the system becomes active, how will it be integrated with other observational networks designed for other purposes? How will it entrain new observations (possibly designed for research purposes or other management goals) as they become available? How might it cope with the loss of measurements as some observational systems are retired?

NEXT STEPS

There are two goals that served as motivation for holding the workshop:

- **Identify a set of indicators to be included in the NCA products that represents the impacts of climate change on ecosystems**. These ecological indicators, along with indicators of changes in the physical climate system and socioeconomic systems, will be included in an indicators chapter.

- **Coordinate among existing observational networks to improve our understanding of the impacts of global change on ecosystems**. To make progress toward this longer-term EIWG goal, a report will be prepared describing the opportunities and challenges of integrating ecosystem observational networks.

These goals have also been an important component of EIWG discussions following the workshop. To achieve these goals, the EIWG envisions two parallel tracks of activities, shown in Figure 4. Both sets of activities would take place between February 2011 and mid-2012. These tracks of activities are designed to address the issues and answer the questions raised in the framework (Figure 2). Ultimately, the tracks should yield draft material that could be incorporated into the 2013 NCA report and other NCA products, as well as a special report regarding the integration of observational networks to improve our understanding of the impacts of global change on ecosystems.

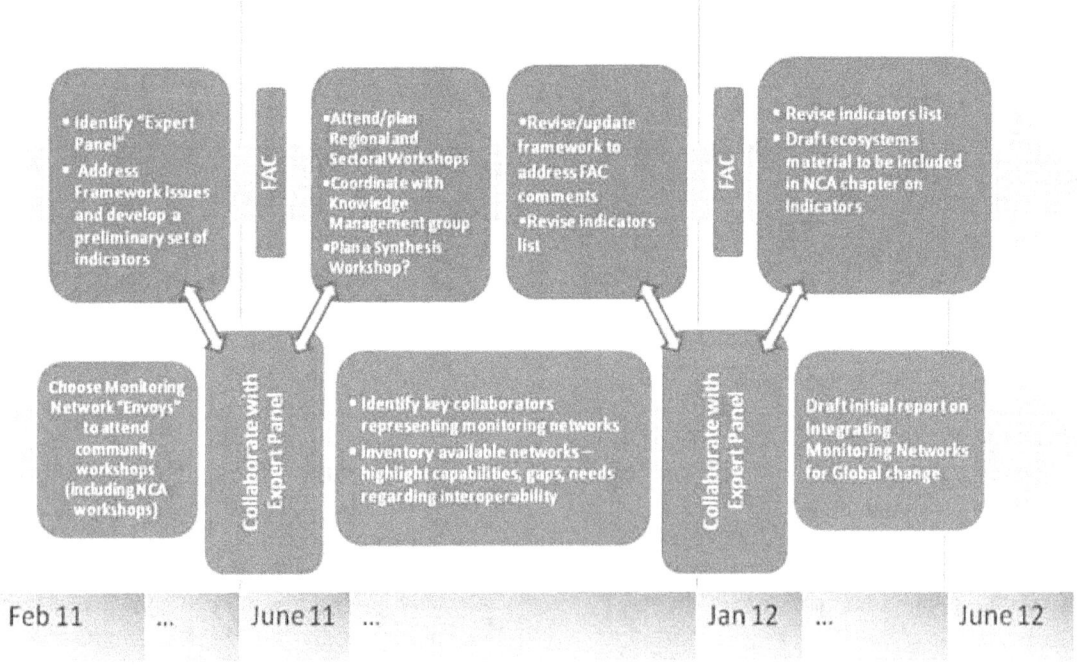

Figure 4: A time line for the indicator selection and monitoring networks integration processes. The activities shown in red (top arrow) contribute to the NCA-related efforts toward preparing a chapter describing ecological indicators. The activities shown in blue (bottom arrow) contribute to the EIWG's longer-term goal to integrate current and future observational networks. Yellow arrows indicate opportunities for the "Envoys" and "Expert Panel" to collaborate, coordinate, and share their progress. "FAC" stands for "Federal Advisory Committee," and represents the steps where a FAC would evaluate material provided by the Expert Panel.

Figure 4 presents an ambitious schedule for the generation of these two products. A diverse collection of experts and stakeholders will need to be involved in these activities, and a minimal amount of time is allotted to holding meetings, writing, and reviewing draft material. The limited amount of time (and resources) will require participants in the process to leverage already-planned activities (either associated with the NCA process or within the scientific and agency communities) and entrain volunteers to assist from across the ecological science community.

Within this relatively short time frame shown in Figure 4, it is unlikely that these processes will yield a comprehensive and operational set of indicators, or integrate multiple observational networks that combine data sets in novel ways (or produce novel data sets). Rather, this plan outlines the important initial steps in a process for achieving those goals, and identifies the appropriate types of people who may be able to accomplish the tasks.

Identification of "Expert Panel" and "Envoys"

The process outlined in Figure 4 relies on the selection of two groups of individuals – an "expert panel" and a set of "envoys" – to do the "heavy lifting" of creating the two intended products for the NCA and the EIWG. Although these two groups could involve some of the same people, the responsibilities of each group will be slightly different, as described in this section. **The expert panel would be tasked with clarifying the Lexicon, Scope/Purpose, and Audience associated with the effort to develop a set of ecosystem indicators** (see Figure 2 and Figure 3). To this end, the expert panel would have to:

- Meet and discuss these issues, identifying important resources (which could include the scientific and policy literature or other experts and speakers) from previous or current efforts to develop indicators.

- Prepare and submit a clarified framework to the NCA's Federal Advisory Committee, the National Climate Assessment Development and Advisory Committee (NCADAC), and then make changes to the framework in response to NCADAC comments.

- Collaborate with those leading other NCA indicator workshops to ensure consistency across the discussions and working groups

that will ultimately contribute to the NCA process and products.

- Collaborate with members of the envoy group (as designated by yellow arrows in Figure 4).

The expert panel should be composed of individuals with an in-depth knowledge of the scientific literature regarding the ecological impacts of climate change, as well as past efforts to select indicators for policy or management purposes. These individuals should have strong connections to leading academic and research institutions and the federal agencies that support ecological research and/or manage ecological resources.

The envoys would act as "ambassadors" for the EIWG, identifying the individuals and organizations that can assist in gathering information about the capabilities and gaps of existing ecosystem observational networks with respect to understanding the impacts of global change. Their ultimate goal will be to generate a report summarizing this information, and analyzing the ways in which existing (or soon-to-be-deployed) observational networks can be integrated or made interoperable. This report should also outline ways in which federal agencies could better coordinate their efforts to observe ecosystems, and to translate their observational data into an improved understanding of the impacts of global change.

Specifically, the envoys would be requested to:

- Attend meetings and workshops where a significant number of experts from the ecological monitoring community, especially those from federal agencies, are expected to participate.

- Attend NCA-related workshops related to indicators, including other USGCRP workshops in the series on Monitoring Climate Change and Its Impacts: Sources for Indicators, Detection, and Attribution as well as workshops for regional and sectoral stakeholders.
- Establish relationships with individuals leading these other meeting and workshop efforts. These individuals could be contacted to serve as contributors or reviewers for future meetings or written reports.

- Collaborate with the expert panel and the EIWG to ensure consistency between observational network-related efforts and those associated with the evolution of the framework and the selection of ecological indicators.

- Organize (with the EIWG as a whole) the process of preparing a report on the integration of ecosystem observational networks.

The envoys would likely hail from a variety of federal agencies, have substantial experience in ecosystem monitoring activities, and represent a range of ecological and climate-related expertise areas (e.g., terrestrial ecology, aquatic/marine ecology, hydrology, climatology).

The expert panel and envoys are likely to be most effective if some members from the EIWG are included in those groups. This would permit the EIWG to have a direct liaison to each group's activities, and to address each group's needs as they arise. For those that are selected to be in either group, a relatively significant time commitment would likely be required during 2011 and 2012.

Incorporating a Wide Range of Input when Selecting Indicators

There are many different individuals and groups (see Figure 3), each with different perspectives and goals that might use a set of ecological indicators. These potential users may adopt indicators to inform decisions, to better understand ecosystems, to engage in basic research, and to shape the science and policy agendas associated with all of these tasks.

Several steps of the process in Figure 4 have been designed to ensure that a diverse set of opinions are reflected in the clarification of the framework for selecting indicators. In choosing the expert panel, the EIWG and NCA staff should include individuals from a broad range of scientific and organizational backgrounds. Also, when planning future meetings and workshops, attendees should hail from a variety

of governmental agencies, academia, the private sector, and the policy-making community. In soliciting and incorporating a wide range of perspectives into the selection of indicators, the process will necessarily be iterative. The details of the framework are likely to be amended and the key questions and answers will be refined over time. It is assumed that this workshop report outlines a starting point, a road map for making progress, and some potential endpoints and products. As the process moves forward, the framework should be amended and the goals and next steps should be appropriately modified.

Communicating to and Coordinating with Other Groups with Similar Goals

The workshop demonstrated that there are a range of activities aimed at integrating existing observational networks (e.g., Climate Effects Network[5], National Aquatic Resource Surveys) or establishing new networks that are designed to incorporate information across many different types of measurements at relatively large spatial scales (e.g., NEON). Trying to entrain the experience and expertise associated with these activities into the EIWG's efforts will be crucial. The time and the resources are not available to "reinvent the wheel" – the EIWG's ability to identify, coordinate with, and build upon the efforts of these other groups will be critical to its success.

The framework outlined in the Section 4 could also be exported to some of the other interagency working groups (IWGs) that are responsible for NCA-related or indicator-related tasks. If the EIWG can communicate this framework effectively to other IWGs, it may provide a valuable source of critical input for the EIWG's work, while easing the burden on other IWGs.

---

[5]It appears that the funding for the Climate Effects Network will not be continued in FY12; it is unclear at this time (May 2011) in what capacity the program's work would be continued.

REFERENCES CITED

Climate Change Science Program (CCSP). (2008). *Our Changing Planet – The U.S. Climate Change Science Program for the Fiscal Year 2009*. Retrieved from http://www.usgcrp.gov/usgcrp/Library/ocp2009/ocp2009.pdf

Environmental Protection Agency (EPA). (2010). *Climate Change Indicators in the United States*. National Service Center for Environmental Publications. Washington, DC. Retrieved from http://www.epa.gov/climatechange/indicators.html

Heinz Center. (2002). *The State of The Nation's Ecosystems: Measuring the Lands, Waters, and Living Resources of the United States*. The H. John Heinz III Center for Science, Economics, and the Environment. Washington, DC.

Heinz Center. (2006). *Filling the Gaps: Priority Data Needs and Key Management Challenges for National Reporting on Ecosystem Condition*. A Report of the Heinz Center State of the Nation's Ecosystem Project. The H. John Heinz Center for Science, Economic and the Environment. Washington, DC.

Heinz Center. (2008). *The State of the Nation's Ecosystems 2008: Measuring the Lands, Waters, and Living Resources of the United States. Highlights*. The H. John Heinz III Center for Science, Economics and the Environment. Washington, DC.

National Research Council (NRC). (2000). *Ecological Indicators for the Nation*. Committee to Evaluate Indicators for Monitoring Aquatic and Terrestrial Environments, National Research Council. Washington, DC.

National Research Council (NRC). (2010). *Monitoring Climate Change: Metrics at the Intersection of the Human and Earth System*s. Committee on Indicators for Understanding Global Climate Change. Washington, DC.

OTHER SOURCES AND REFERENCES

European Environment Agency (EEA). (2008). *Impacts of Europe's changing climate – 2008 indicator-based assessment*. EEA Report No. 4/2008. DOI 10.2800/48117. http://www.eea.europa.eu/publications/eea_report_2008_4

Global Climate Observing System (GCOS). (2010). *Implementation Plan for the Global Observing System for Climate in Support of the UNFCCC (2010 Update)*. United Nations Environment Programme and International Council for Science. Geneva, Switzerland. Retrieved from http://www.wmo.int/pages/prog/gcos/Publications/gcos-138.pdf

International Geosphere-Biosphere Programme. (2009). IGBP Climate-Change Index, from http://www.igbp.kva.se/page.php?pid=504

Jones, K. B., Bogena, H., Vereecken, H., & Weltzin, J. F. (2010). Design and Importance of Multi-tiered Ecological Monitoring Networks. In F. Müller, C. Baessler, H. Schubert & S. Klotz (Eds.), *Long-Term Ecological Research: Between Theory and Application* (Vol. XVII, pp. 355-374): Springer Science+Business Media.

Keller, M. (2010). *NEON Scientific Data Products Catalog*. National Ecological Observatory Network.

Millennium Ecosystem Assessment. (2005). *Ecosystems and Human Well-being: Current State and Trends*. Findings of the Condition and Trends Group of the Millennium Ecosystem Assessment. Washington, Covelo, London.

National Aeronautics and Space Administration (NASA). (2010). Global Climate Change: Key Indicators Retrieved from http://climate.nasa.gov/keyIndicators/

National Climatic Data Center (NCDC). (2010). Global Climate Change Indicators Retrieved from http://www.ncdc.noaa.gov/indicators/

National Oceangraphic and Atmospheric Administration, Arctic Research Office. (2010). Arctic Change: A Near-Realtime Arctic Change Indicator Website, from http://www.arctic.noaa.gov/detect/index.shtml

National Science and Technology Council (NSTC). (1997). *Integrating the Nation's Environmentla Monitoring and Research Networks and Program*s: A Proposed Framework. The Environmental Monitoring Team, Committee on Environment and Natural Resources, National Science and Technology Council. Washington, DC.

Office of Environmental Health Hazard Assessment (OEHHA). (2009). *Indicators of Climate Change in California*. Sacramento, CA.

APPENDIX A: INDICATOR LISTS

Components or Examples		NRC 2000	Heinz	EEA	NRC 2010	CA	EPA	NOAA	NASA	IUCN
Terrestrial										
Land Cover	Forest specific (CA, EEA)	✓	✓	✓	✓	✓				
Land Use		✓	✓		✓					
Productivity	Production Capacity (Chlorophyll per unit area), Net Primary Productivity, Carbon Storage (NEP)	✓								
Species Diversity	Total and Native (NRC 2000); alpine and subalpine (CA and EEA)	✓	✓	✓	✓					
Carbon storage		✓	✓							
Soil Organic Matter		✓	✓	✓						
Soil Erosion			✓	✓						
Soil Moisture			✓	✓	✓					
Animal Migration	Bird Wintering Ranges (EPA), Bird arrival timing (CA), small mammals (CA), Butterfly timing (CA)					✓	✓			
Phenology changes	Leaf and Bloom Dates for lilac and honeysuckle (EPA), Wine grape (CA); various plants and animals (EEA)			✓		✓	✓			
Disturbances	Wildfire		✓	✓	✓	✓				
CO$_2$ fertilization monitoring	Time series of Leaf Area Index				✓					
Albedo change					✓					
Invasive species			✓							
Indicator/Sensitive Species (Range, Abundance, or Changes in inter-species relationships)	Tree mortality (CA), alpine communities (EEA); latitudinal shifts in species ranges (EEA); changes in butterfly-host flower relationships (EEA)		✓	✓	✓	✓				
Extinction Risk										✓
Agriculture Focus										
Nutrient Use Efficiency		✓	✓							
Nutrient Balance	"Inputs and Outputs" (Heinz)	✓	✓		✓					
Crop Timing				✓						
Crop Yield			✓	✓	✓					
Livestock populations					✓					
Water balance			✓	✓	✓					

Components or Examples		NRC 2000	Heinz	EEA	NRC 2010	CA	EPA	NOAA	NASA	IUCN
Freshwater										
Nutrient Runoff		✓	✓		✓					
Lake Trophic Status	Secchi-disk transparency, total phosphorus, and chlorophyll a	✓								
Connectivity			✓							
Acidity			✓							
Stream Oxygen		✓								
Water Quality	Various biological or chemical measures (EEA)		✓	✓	✓					
Water Temperature	Lake Tahoe, Delta specific (CA)			✓	✓	✓				
Habitat Quality			✓							
Species Diversity	Total and Native (NRC 2000)	✓	✓	✓						
Indicator/ Sensitive Species (Abundance)			✓							
Marine and Coastal										
Nutrient Runoff	(only coastal considered, NRC)	✓	✓							
Productivity	(only coastal considered, NRC)	✓			✓					
Species Diversity	Total and Native (only coastal considered, NRC)	✓	✓		✓					
Water Temperature	SST		✓	✓	✓					
Dissolved Oxygen	CA current		✓		✓	✓				
Coastal Erosion			✓		✓					
Shoreline type			✓		✓					
Disturbances	Algal blooms, Hypoxia		✓		✓					
Ocean Color	Also Chlorophyll concentration		✓	✓						
Acidification					✓					
Indicator/Sensitive Species (Range and/or Abundance)	Copepod and Cassin's auklet (CA); zooplankton, tropical fish, decapods (EEA); effectiveness of marine protected areas		✓	✓	✓	✓				

Components or Examples		NRC 2000	Heinz	EEA	NRC 2010	CA	EPA	NOAA	NASA	IUCN
Other Physical Climate Indicators										
Air Temperature	Length of Growing Season, Plant Hardiness Zones, Extremes or Mean (Annual/ Seasonal)			✓	✓	✓	✓	✓	✓	
Precipitation	Mean (annual/seasonal) or extremes			✓	✓	✓	✓			
Snow, Snowmelt, Glaciers, Sea ice, and Permafrost	Runoff, SWE, Snow cover, Glacier area, Glacier volume, Greenland Ice Sheet monitoring, Arctic sea ice monitoring, lake/ stream ice cover dates			✓	✓	✓	✓	✓	✓	
Ocean heat content					✓		✓	✓		
Stability of ocean circulation					✓					
Sea Level				✓	✓	✓	✓	✓	✓	
Storm Surge				✓	✓					
Tropospheric Ozone				✓	✓					
Streamflow	Low flows and floods	✓		✓	✓					
Groundwater depth		✓			✓					
Tropical storm frequency and intensity					✓					
Cloud properties					✓					

NRC 2000 - *Ecological Indicators for the Nation*
Heinz - *The State of the Nation's Ecosystems 2008: Measuring the Lands, Waters, and Living Resources of the United States* (Heinz Center, 2008)
EEA - *Impacts of Europe's changing climate - 2008 indicator-based assessment* (European Environmental Agency, 2008)
NRC 2010 - *Monitoring Climate Change: Metrics at the Intersection of the Human and Earth Systems* (National Research Council, 2000)
CA - *Indicators of Climate Change in California* (Office of Environmental Health Hazard Assessment, 2009)
EPA - *Climate Change Indicators in the United States* (EPA, 2010)
NOAA - *Global Climate Change Indicators*
NASA - *Key Indicators of Global Climate Change*
IUCN - *International Union for Conservation of Nature*, as described in Hoffman et al., 2010,
The Impact of Conservation on the Status of the World's Vertebrates, *Science*, **330**, 1503-1509

URL links appear at https://sites.google.com/a/usgcrp.gov/eco-monitoring/

APPENDIX B: INDICATOR SELECTION CRITERIA

REPORT	Conceptual/ Theoretical	Practical/ Technical/ Data	Other	Limitations in applying to climate change or workshop goals	Links
NRC 2000	Importance Conceptual Basis Reliability Temporal and Spatial Scales of Applicability Statistical Properties Robustness	Data Required and Available Data Archives Data Quality Robustness Skills Required to Collect Data	International Compatility Cots, Benefits, Cost-Effectiveness	Not designed for CC, but for overall ecological monitoring	http://books.nap.edu/catalog.php?record_id=9720
Heinz 2008	National Scale	Data availability (now or in the future) Data credibility Trends, Reference points, and regional disaggregation preferred	Designed for policy makers and opinion leaders. Many indicators do not currently have data; designed to highlight gaps/priorities	Focuses on the states of ecosystems (rather than pressures/ processes)	http://www.heinzctr.org/publications/
EEA 2008	Causal link to climate change Policy Relevance	Measurability Availability of historical time series (in many cases, 20 years or more) Data availability over Europe Transparency (clear to policymakers and public)		Data and analyses taken mostly from scientific literature (rather than an accessible integrated network of indicators)	http://www.eea.europa.eu/publications/eea_report_2008_4
NRC 2010	Direct connection to climate change (or related impact processes)Significance Dominance (relative to other factors)	Measurable Historical Well documented	Each category (Ocean, Terrestrial, Cryosphere, Atmosphere, Hydrology, Disasters) has its own selection criteria	Ultimately focused on impacts to human systems and well-being	http://books.nap.edu/catalog.php?record_id=12965
CA 2009	Follow IPCC model linking climate drivers -> climate change -> climate impacts Indicators are sensitive to climate change Linked to Decision Making	Data Quality	Data identification and classification of quality follow Environmental Protection Indicators for California (EPIC) standards, which apply to many other environmental and health indicator/monitoring programs	Often site/resource specific	http://www.oehha.ca.gov/multimedia/epic/climateindicators.html
EPA 2010	Relevance to climate change Ability to show a meaningful trend	Usefulness Objectivity Data Quality Transparency		Ecosystem indicators are often derived from air temperature data (e.g., growing season = time between spring and fall frost)	http://www.epa.gov/climatechange/indicators/pdfs/ClimateIndicators_full.pdf